双语版·全4册

BUMBLEBEE

亲亲动物

如果蜜蜂来我家

[英]弗朗西斯·罗杰斯 [英]本·克里斯戴尔 著绘 范晓星 译 朱朝东 丁亮 审校

中信出版集团 | 北京

图书在版编目（CIP）数据

如果蜜蜂来我家 : 汉文、英文 / (英) 弗朗西斯·罗杰斯, (英) 本·克里斯戴尔著绘 ; 范晓星译. -- 北京 : 中信出版社, 2023.3
（DK亲亲动物 : 双语版 : 全4册）
ISBN 978-7-5217-5239-7

Ⅰ. ①如… Ⅱ. ①弗… ②本… ③范… Ⅲ. ①蜜蜂－少儿读物－汉、英 Ⅳ. ①Q969.557.7-49

中国国家版本馆CIP数据核字（2023）第021852号

Original Title: How can I help Rosy the bumblebee?

如果蜜蜂来我家
（DK 亲亲动物双语版　全 4 册）

著　　绘：［英］弗朗西斯・罗杰斯　［英］本・克里斯戴尔
译　　者：范晓星
出版发行：中信出版集团股份有限公司
（北京市朝阳区东三环北路 27 号嘉铭中心　邮编　100020）
承 印 者：北京顶佳世纪印刷有限公司

开　　本：889mm × 1194mm　1/20　　印　　张：2　　字　　数：115 千字
版　　次：2023 年 3 月第 1 版　　印　　次：2023 年 3 月第 1 次印刷
京权图字：01-2022-4478　　审 图 号：GS 京（2022）1525号
书　　号：ISBN 978-7-5217-5239-7
定　　价：156.00 元（全 4 册）

出　　品　中信儿童书店
图书策划　红披风
策划编辑　陈瑜
责任编辑　袁慧
营销编辑　易晓倩　李鑫橦　高铭霞
装帧设计　哈_哈

如有印刷、装订问题，本公司负责调换。
服务热线：400-600-8099
投稿邮箱：author@citicpub.com

致所有好奇的孩子！

For the curious
www.dk.com

混合产品
纸张｜支持负责任林业
FSC
www.fsc.org
FSC® C018179

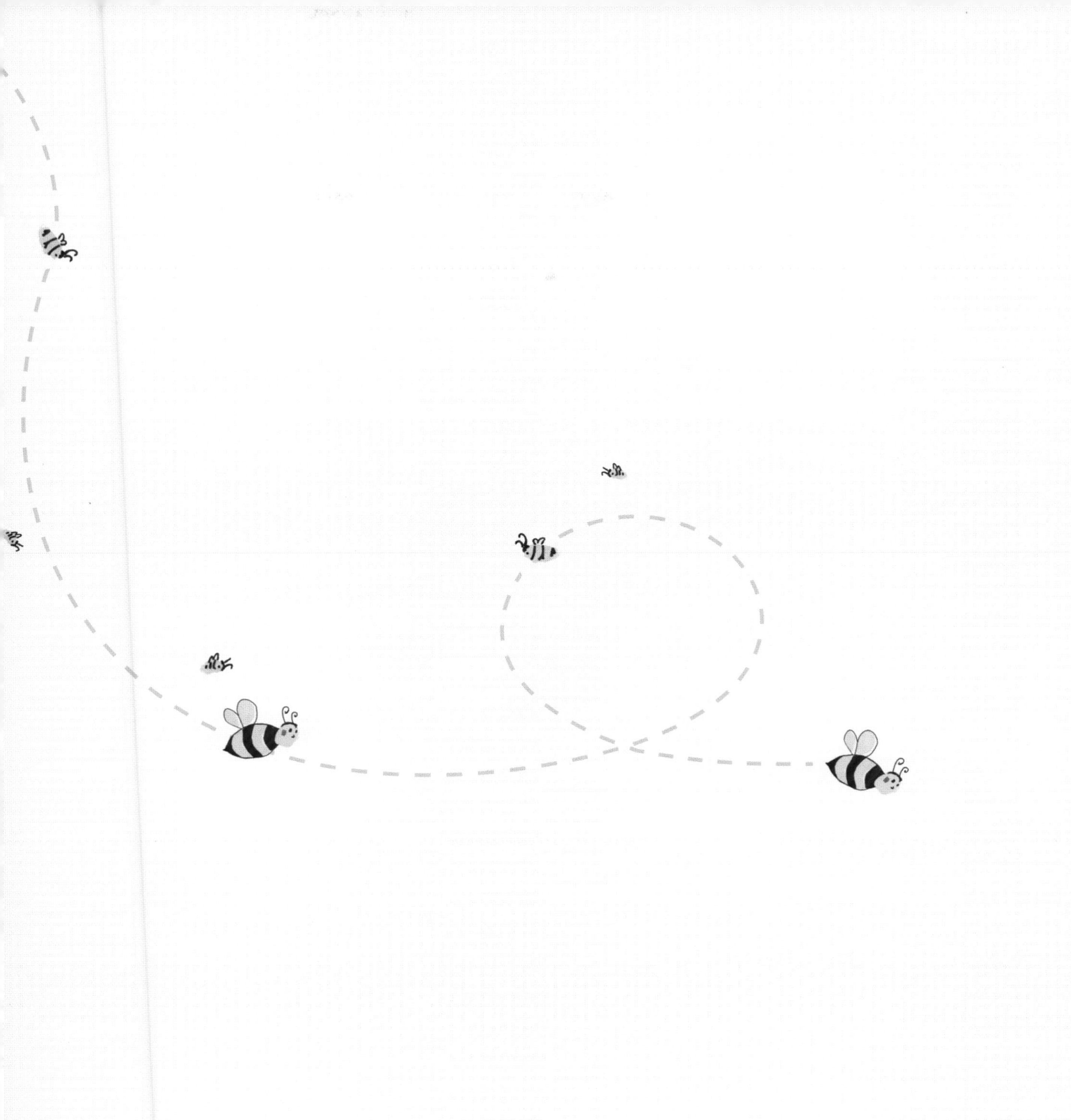

你好，我叫罗丝。

我是一只熊蜂。

我想请你帮帮忙，可以吗？

Hello, my name is Rosy.

I am a bumblebee and I need your help.

我想到你家花园做客。

I like to visit your garden.

花 园
GARDEN

花蜜
Nectar

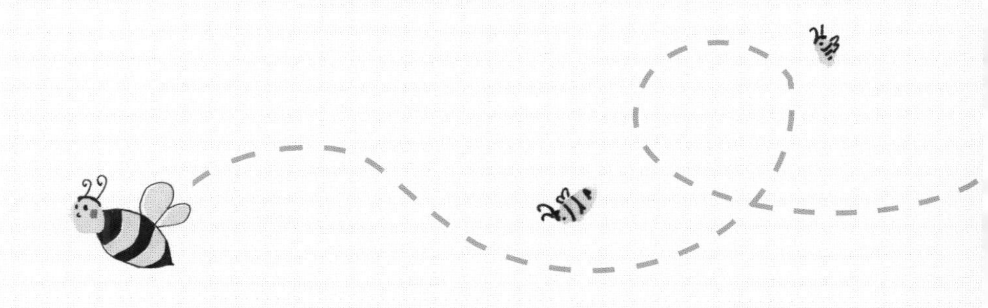

我爱吃花粉和花蜜。

I like to feed on the
pollen and nectar from your flowers.

请在你的花园里
为我种很多很多花吧。

Please plant lots of flowers
in your garden for me.

即便你家没有很大的院子，
也可以在花盆或者吊篮里，为我养一些花。

Even if you do not have a lot of outdoor space,
you can plant flowers in a pot or basket.

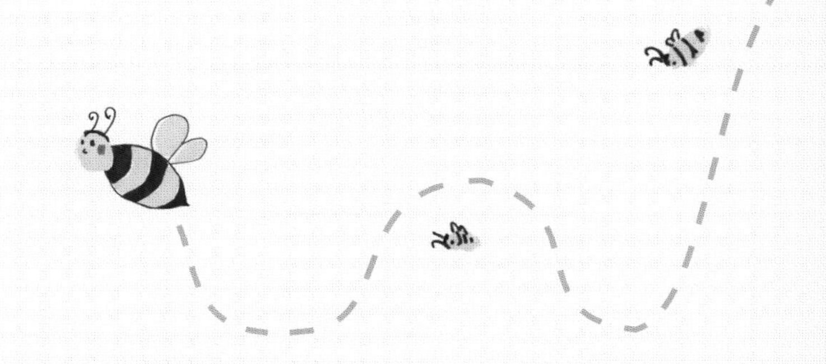

我找花的时候很辛苦，
会非常渴。

I get really thirsty when I am
looking for flowers.

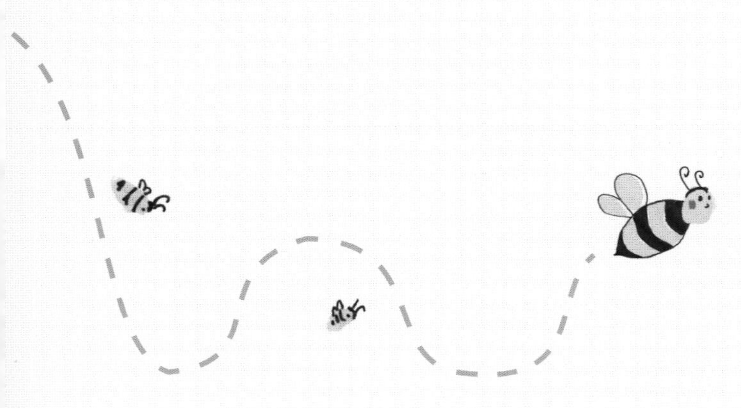

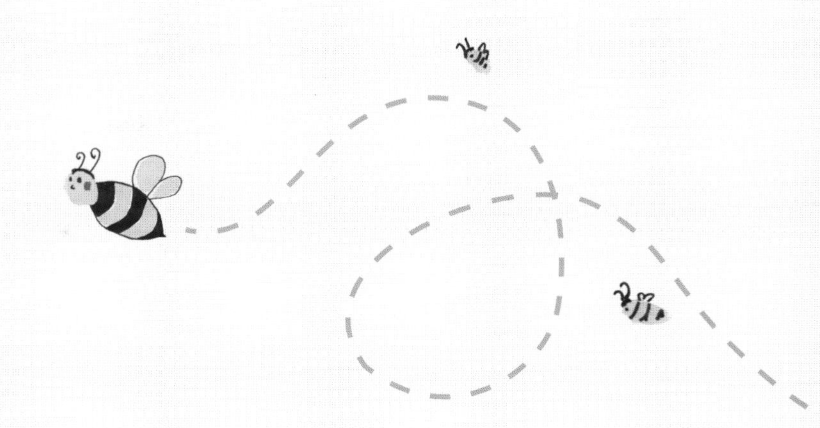

请给我留一些水吧。
但盛水的容器可不要太深哟！

Please leave some water out for me.
But make sure it is not too deep!

我喜欢在树上安家、睡觉。

I like to nest and

sleep in trees.

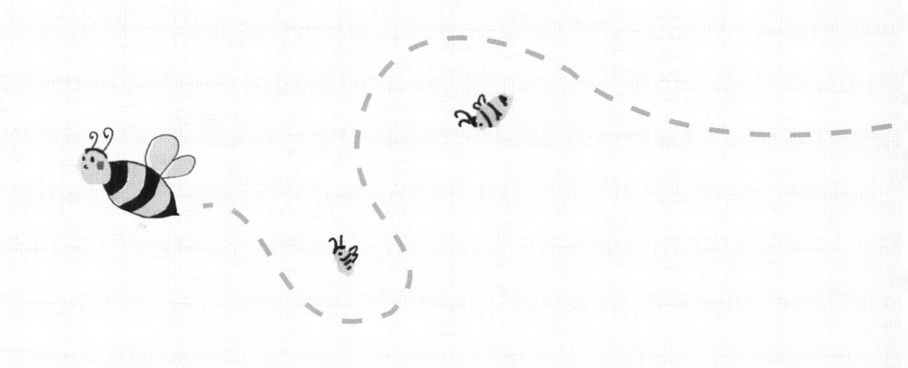

在你家花园里为我种几棵树吧。

Please plant some trees
for me in your garden.

我需要温暖、安全的地方休息。

I need somewhere warm
and safe to stay.

建一座蜜蜂类昆虫旅馆是个好主意。
我在里面会很暖和，
就算下雨也不怕了。

Please make me a bee hotel to live in.
It will keep me warm and protect me from the rain.

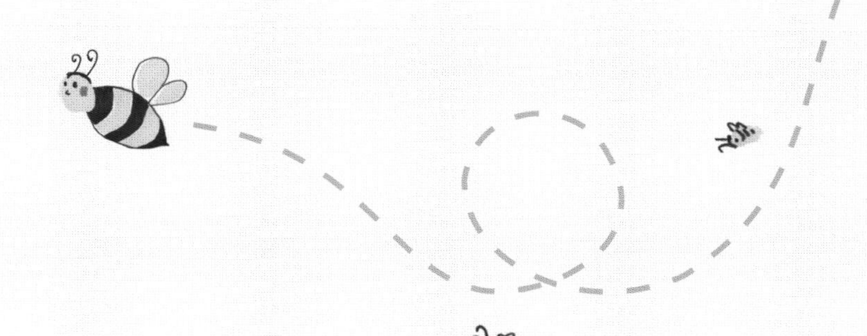

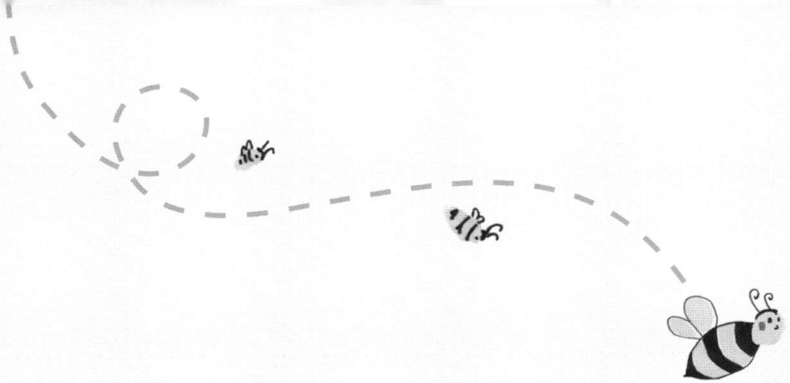

我每天飞来飞去，太累啦，
需要休息的地方。

I get tired from flying all day
and need somewhere to rest.

噜

呼

噜

呼

噜
呼
噜
呼

请在你的花园里
布置一个自然角，
我就有睡觉的地方了。

Please make a wild corner
in your garden with
somewhere for me to sleep.

谢谢你为了保护我们做的一切。

Thank you for all your help.

我们为什么要保护熊蜂？

Why do we need to protect bumblebees?

罗丝是一种叫熊蜂的蜜蜂。熊蜂在世界很多地方都可以看到，它们经常出现在花园、草地、果园、树林和公园里。遗憾的是，熊蜂现在面临生存危机。

Rosy is a type of bee called a bumblebee. Bumblebees live all around the world and can be found in gardens, meadows, orchards, woods, and parks. Unfortunately, these bees are in danger.

我需要你的帮助！
I need your help!

在一些地方，气候变得炎热，熊蜂无法生存，因此它们濒临灭绝（在地球上再也见不到它们了）。

In some places it is getting too hot for bees to survive, so they are at risk from becoming extinct (no longer existing).

罗丝这样的熊蜂小小的、毛茸茸的，那么可爱，我们要尽力去保护它们。

We need to keep these tiny, fluffy bumblebees, like Rosy, safe and do our best to protect them.

传粉

Pollination

蜜蜂很重要，它们帮助传粉。花朵需要传粉，这样才能长出种子
繁殖出下一代。花粉可以通过风或者像罗丝这样的动物等传播。

Bees are important as they help with pollination. Flowers need pollination
make seeds that then grow into new plants. Pollen can be moved by the win
or by animals, like Rosy.

熊蜂落在花上，吃又香又
甜的花蜜（含糖的液体）

Bumblebees land on flowe
to eat the delicious, swe
nectar (a sugary liquid).

花上的花粉会沾在熊蜂的触角、脸和身体上。蜜蜂将花粉储存在花粉篮里，带回蜂巢。

The pollen from flowers then gets stuck on them. It sticks to their antennae, face, and body. Bees store pollen in pollen baskets to take back to the hive.

当蜜蜂落在其他花朵上面的时候，花粉从它们的身上掉落。

The pollen then falls off bees when they visit the next flower.

花粉篮
Pollen basket
没有我们，好吃的水果和坚果就会变少了，比如橙子、草莓和扁桃！
Without us you would have less tasty fruit and nuts, such as oranges, strawberries, or almonds to eat.

蜜蜂的种类

Different types of bees

罗丝是熊蜂。世界上大约有 2 万种蜜蜂。从小小的、会跳舞、会采集花蜜的蜜蜂，到只有用放大镜才能看到的非常非常小的蜜蜂！你见过哪些蜜蜂呢？

Rosy is a bumblebee but there are around 20,000 different types of bees found across the world. From small, dancing bees that collect honey, to teeny tiny ones that you would only see if you used a magnifying glass! Which bees have you seen?

蜜蜂和马蜂的区别

What is the difference between a bee and a wasp?

蜂和马蜂是很难分辨的，因为它们看起来很像。但它们是有区别的。当你花园里寻找蜜蜂的时候，可以注意以下几点。

an be tricky to tell the difference between a bee and a wasp because they look so similar. t they have differences, too. re are some things to watch t for when you are looking bees in your garden…

马蜂的触角弯弯的，蜜蜂的触角更长、更细。

Wasps have curled feelers, but bees' antennae are longer and thinner.

蜜蜂

Bee

马蜂

Wasp

蜂有一些绒毛，而蜜体表都是很密的绒毛。

asps have some hair, t bees have hair all er their body.

通常它们都有六条腿。

Usually, both have six legs.

蜂的身体看起来比马蜂要圆，蜂的身体细长，所以它们飞得快。

es tend to look rounder than asps. Wasps usually have long d slim bodies to help them fly ster.

马蜂身上的条纹通常比蜜蜂多。

Wasps usually have more stripes on their bodies than bees do.

蜂的颜色大多是橘黄色和黑色相间，马蜂大多是黄黑色。

es look more orange and black in colour, whereas asps are bright yellow and black.

马蜂的刺更光滑，蜜蜂的刺有很多小小的倒钩。

Wasps' stingers are smooth, whereas bees' stingers are barbed and have lots of little hooks on them.

喜欢的花

Favourite flowers

熊蜂喜欢漂亮、五颜六色的花。在你家花园里多种些花吧，看看有多少小蜜蜂会来你家做客。

Bumblebees are attracted to a range of beautiful and colourful flowers. Try planting some of these plants in your garden, and see how many bees you can spot...

毛地黄
Foxglove
黄番红花
Yellow crocus
红花苜蓿
Red clover
琉璃苣
Borage
聚合草花
Comfrey
薰衣草
Lavender
报春花
Primrose
黑矢车菊
Black knapweed

鸣 谢

本出版社感谢以下机构提供照片使用权：

（a= 上方；b= 下方；c= 中间；f= 底图；l= 左侧；r= 右侧；t= 顶端）

33 123RF.com: Warut Chinsai / joey333 (tr). **35 Dreamstime. com:** Sander Meertins (cla); Krzysztof Slusarczyk (br). **36 123RF. com:** paulrommer (cra). **Dorling Kindersley:** Neil Fletcher (ca); Jacinto Valter (c). **36–37 Dreamstime. com:** Tirachard Kumtanom. **38 Dreamstime.com:** Photographieundmehr (ca). **39 123RF. com:** Uliana Dementieva (tc). **Dreamstime.com:** Krzysztof Slusarczyk (br). **40 Dreamstime.com:** Krzysztof Slusarczyk (crb).

其余图片版权归英国 DK 公司所有，更多信息请访问 www.dkimages.com。

说明：传粉是指花粉从一朵花传到另外一朵花（传粉者将雄蕊花药里的花粉传到雌蕊的柱头或胚珠上）。

关于作者和绘者

本和弗朗西斯是一对夫妻，住在英国恩河畔纽卡斯尔。他们倾心于帮助家中花里的野生动物。仲夏的夜晚，弗朗西斯醒有了一个灵感——创作鼓励小朋友加入他的行动的系列童书。

弗朗西斯创作故事，本画插图，他们领小读者进入一个栩栩如生的野生动物世快来欢迎花园里的小客人：小刺猬罗里、麻雀罗芮、小蜜蜂罗丝、小蝴蝶罗克西。